Nthulane Makgato

Key Performance Indicators analysis for the SAPS Vehicle Depot

AF153251

Nthulane Makgato

Key Performance Indicators analysis for the SAPS Vehicle Depot

Establishing performance measures and developing balanced scorecard

LAP LAMBERT Academic Publishing

Impressum / Imprint

Bibliografische Information der Deutschen Nationalbibliothek: Die Deutsche Nationalbibliothek verzeichnet diese Publikation in der Deutschen Nationalbibliografie; detaillierte bibliografische Daten sind im Internet über http://dnb.d-nb.de abrufbar.
Alle in diesem Buch genannten Marken und Produktnamen unterliegen warenzeichen-, marken- oder patentrechtlichem Schutz bzw. sind Warenzeichen oder eingetragene Warenzeichen der jeweiligen Inhaber. Die Wiedergabe von Marken, Produktnamen, Gebrauchsnamen, Handelsnamen, Warenbezeichnungen u.s.w. in diesem Werk berechtigt auch ohne besondere Kennzeichnung nicht zu der Annahme, dass solche Namen im Sinne der Warenzeichen- und Markenschutzgesetzgebung als frei zu betrachten wären und daher von jedermann benutzt werden dürften.

Bibliographic information published by the Deutsche Nationalbibliothek: The Deutsche Nationalbibliothek lists this publication in the Deutsche Nationalbibliografie; detailed bibliographic data are available in the Internet at http://dnb.d-nb.de.
Any brand names and product names mentioned in this book are subject to trademark, brand or patent protection and are trademarks or registered trademarks of their respective holders. The use of brand names, product names, common names, trade names, product descriptions etc. even without a particular marking in this works is in no way to be construed to mean that such names may be regarded as unrestricted in respect of trademark and brand protection legislation and could thus be used by anyone.

Coverbild / Cover image: www.ingimage.com

Verlag / Publisher:
LAP LAMBERT Academic Publishing
ist ein Imprint der / is a trademark of
AV Akademikerverlag GmbH & Co. KG
Heinrich-Böcking-Str. 6-8, 66121 Saarbrücken, Deutschland / Germany
Email: info@lap-publishing.com

Herstellung: siehe letzte Seite /
Printed at: see last page
ISBN: 978-3-659-40235-7

Acknowledgements

- I would like to thank my Lord and Savvior Jesus Christ. He is my Beginning and my End. I am nothing without him.
- My family. You have shown me great support over the years and I'm grateful that God chose to put you in my life. I cannot imagine how I could be here without you.
- South African Police Service vehicle depot. For allowing me to base my final year project on you. You were always reading and willing to help and I greatly appreciate that.
- Fourier Approach. Without vacation work at your company, I would not have had this opportunity. Thank you for your input in the project and guiding me to its completion.
- My project leader. Dr Antonie van Rensburg, your high standards and insights were invaluable.

Executive Summary

Police service is one of the most critical in a community and society at large. To limit crime and ensure that the law is upheld, resources need to be readily available to the South African Police Service (SAPS). A full fleet of well-maintained police vehicles is one such crucial resource. Depots performing regular maintenance services and repairs and panel beating are available to the SAPS. Despite the availability of these depots in various regions, vehicles booked in for services spend a prolonged time at the depot. The longer a vehicle is at the depot, the less efficient it is in crime prevention. One such SAPS depot can be found in the Silverton area. This report focuses on the Silverton depot where a challenge contributing to under-performance has been identified: a lack of established Key Performance Indicators.

Key Performance Indicators (KPI) are one of the most important aspects of an organization as they reflect how the company is performing. Even inaccurate information regarding KPIs is dangerous due to the fact that critical decisions would be made based on this incorrect information. KPIs and the manner in which they are measured for the SAPS are thus all the more important as it is one of the basic and very key services that are provided in this country.

This document explains how the student will explore ways in which one can improve the state of the depots in terms of their performance measures through the use of Industrial Engineering principles such as surveys, research and the Balanced Scorecard.

Table of Contents

List of Figures

List of Tables

1. Introduction and Background

The South African Police Service has come a long way since the era of apartheid where there were 10 homelands known as the TBV states. These included Transkei, Bophutatswana, Venda and Ciskei and the "Self-governing territories" included Kwazula, Lebowa, Qwaqwa, Gazankulu and Kwandebele. Each homeland had its own policing agents which meant that each one had its respective uniform, rank and was established on its own section of legislation. Before 1995 there were 11 different policing agencies – 10 homelands and 1 South African Police Service. After Nelson Mandela was elected President, he appointed George Fivaz as the first National Police Commissioner of the SAPS, it became his responsibility to unite the 11 policing agencies into one SAPS.

The SAPS has a number of items that are at the core of its functions which include Visible Policing, Protection and Security Service, Crime Intelligence, Criminal record and Forensic Science Service. All these service are important in providing safety and protection to South Africans (SAPS Profile, 2011).

Amongst other things, the SAPS require quality and efficient mobility in order to perform its functions. Emergency responses, police travel and escorts require a fleet in its best condition. The South African Police Service has a number of vehicles on the roads of Gauteng. The vehicles serve the South African public through crime prevention. Wear and tear as well as accidents occur naturally and circumstantially. Police vehicles are therefore serviced, repaired and panel beaten on a regular basis at depots throughout the country. Regrettably, however, many of these vehicles are not being utilised in combating crime as they should, but are instead waiting to be processed at the depots.

The depot that is the focus of this project was founded in the Pretoria Central Business District (CBD) in 1913 on the corner of Pretorius and Shubert Street. In 1980 the depot was moved from the CBD to Silverton on Moreleta Street.

The SAPS divides a region into clusters that must be monitored by certain stations. A station's cluster is their jurisdiction. The depots have a number of clusters that are assigned to them. All the vehicles belonging to the stations in the assigned cluster, therefore, will be repaired at the depot. The Silverton depot is home to the Sunnyside cluster (stations: Sunnyside, Brooklyn, Garsfontein, Lyttleton, Pretoria Moot and Villeria), Pretoria Centralcluster (stations: PTA central, Atteridgeville, Erasmia, Hercules, Laudium, Pretoria

1

West,Wierdaburg, Wonderboompoort) and the Mamelodi cluster (stations: Mamelodi East, Mamelodi West, Eersterust, Kameeldrift, Silverton, Sinoville).

This project focuses on assisting the Pretoria depot in informing it of its performance through Balanced Scorecard and therefore improving the depot efficiency. Greater efficiency translates in more police vehicles back on the roads performing the functions which are most imperative to communities.

2. Depot Environment

The garage is a large facility, which is not surprising considering all the types of services that they perform and the amount of stations that bring their cars on an almost weekly basis. The size of the facility is estimated at 790m^2. The following diagram shows the aerial view of the layout of the facility and the path of its processes.

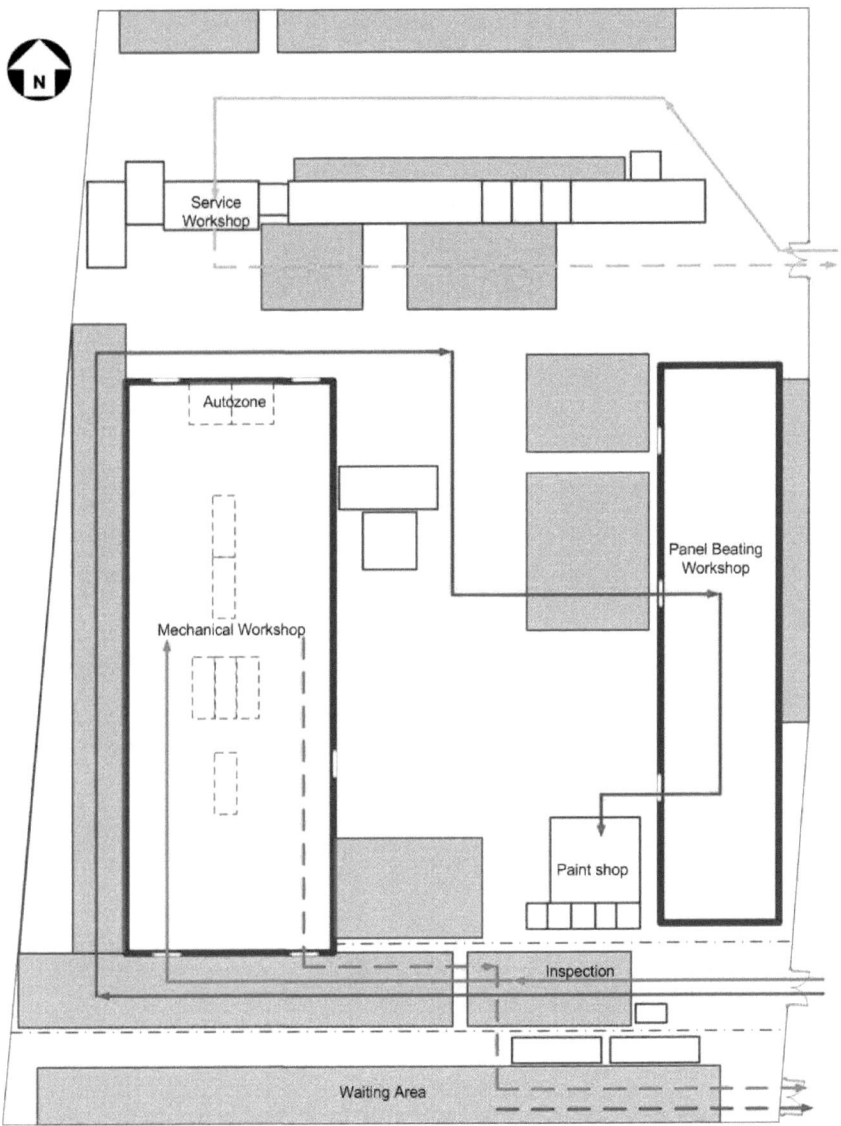

Figure 1: Aerial view of the layout of the depot

Figure 1 shows the aerial view of the layout of the facility. It also shows the important areas in the layout of the facility, these are the mechanical workshop, service workshop, panel beating workshop, Autozone, the inspection area and the waiting area. The diagram also shows the process flow for the different processes that are done at the depot. The green line shows the flow for mechanical repair, the yellow line is for service and the red line is for panel beating. The dashed lines for all 3 processes shows the path after the vehicle has been processed and is leaving the depot.

Figure 2 below shows a typical process map for what takes place when the service of a vehicle takes place.

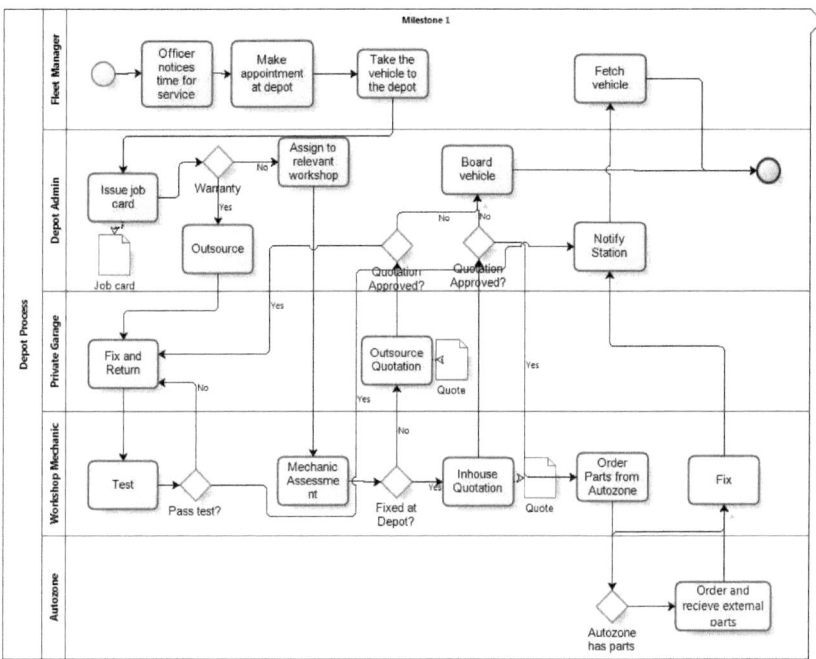

Figure 2: Process Map for service, repair and panel beating

3. Project Problem

Vehicles at the South African Police Service depot are not being processed at a satisfactory rate and one of the challenges that contribute to this problem is the lack of performance measures.

It is important that the rate of vehicle processing be increased as the vehicles play a very important role in the battle against crime. Crime statistics for Gauteng from April 2009 to March 2010 from the SAPS website show that the following crimes occurred:

Nature of Crime	Statistic
Sexual crimes	15 645
Common robbery	20 107
Theft of vehicle or motor vehicle	36 337

Table 1: Crime Statistics

These numbers alone indicate the dire state of crime in Gauteng even without taking murder, attempted murder, kidnapping, drug-related crime etc., into consideration. The depot has an average weekly demand of 453 vehicles that it needs to process, it processes 279 vehicles and it carries over 174 vehicles and these vehicles belong to the clusters of the stations that fight crime. The statistics show the importance of having police officers that are fully equipped with fleet in good condition so as to prevent and combat crime as far as possible.

This is consistent with a study conducted in 1987 by the National Association of Accountants and Computer Aided Manufacturing International. The survey put forward that of 260 financial officers and 64 executives, approximately 60% were not satisfied with their performance measures. At a later stage, a similar survey was taken and 80% of large American companies were not satisfied with their performance measures and wanted to change (Niven, 2002).

The Silverton depot is facing a number of challenges that stop it from processing the number of vehicles that are required by the stations. Some of their challenges are documented in the diagram below.

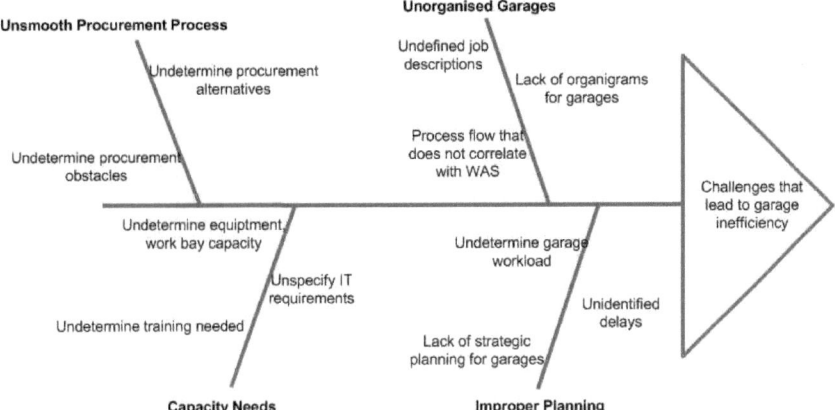

Figure 3: Fishbone of challenges at the depot

4. Project Aim and Objectives

One cannot manage what one cannot measure. Performance measures allow one to know how well or how poorly one is faring in a particular area. The benefit of consistent performance measurement is that management is based on accurate information. Decisions that are founded on accurate information will be reliable. This information is very important in the decision making process particularly when one makes improvements or changes.

The aim of the project is to define the appropriate performance indicators relevant to the depot, how they are measured and how they should be utilised and managed. The project endeavors to

- Identify performance indicators.
- Recommend a manner in which to measure the performance measures.
- Create a balanced scorecard.

5. Project Scope

This project will only focus on performance measures that are relevant to the police depot in Silverton Pretoria. The focus will be on automotive services performed on mechanical repairs for breakdowns and post-accident panel beating. There are several other processes that take place at the depot but these will not be included in this scope. Albeit the information for this project is focused specifically on the Pretoria depot; however it may be relevant to the rest of the Gauteng region.

6. Deliverables

The depot in Silverton currently does not have an established performance measurement system. This project will create a more consistent and centralised measurement system through the Balanced Scorecard.

The following deliverables have been defined for this project:

- Identified performance measures.
- Developed measuring techniques.
- Design of an interface that displays the performance of the depot through a balanced scorecard.
- Recommendations on the management of the performance indicators.

7. Literature Review

In the book *Private Muscle*, Anthony Minaar and DuxitaMistry state that, "After South Africa's first democratic election in 1994, levels of recorded crime increased substantially. Public fear of crime soared, as did the perception that criminals were breaking the law with impunity. Public perception was driven by the belief that the country was facing a 'crime explosion'. Crimes such as vehicle hijacking, rape and murder received prominent police coverage." Acts such as these have since increased the importance of police presence and activity throughout the country (Minaar and Mistry).

Johan Burger analyzes the rate of response of the SAPS by looking at the 10111 call centers (receiving emergency calls) and the Flying Squad (responding to emergency calls). In his article, he mentions one of the greatest problems mentioned by many a fleet manager regarding the SAPS garages, "The 10111 centers and flying squads are often accused of poor performance and ineffectiveness. The Pretoria News reported on 17 February 2009 that only four out of 24 vehicles of the Pretoria flying squad were fit for use. This means that of the 12 to 15 members per shift, only eight will be operationally active, and only four as opposed to six or seven vehicles can be deployed at any given time. According to the Pretoria News report, the flying squad blames the 'unserviceability' of many of its vehicles on incompetency at the police garage"(Burger, 2009).

Burger's statement highlights a critical issue found at the SAPS depot/garage, the challenge of managing one's assets. According to the United States Department of Transportation, "Asset management is a systematic process of maintaining, upgrading, and operating physical assets cost-effectively. It combines engineering principles with sound business practices and economic theory, and it provides tools to facilitate a more organized, logical approach to decision-making. Thus, asset management provides a framework for handling both short- and long-range planning." The department also goes on to say that an asset management system should be customer focused, mission driven, system orientated, long-term in outlook, accessible, user friendly and flexible (U.S. Department of Transport, 1999).

7.1. Balanced Scorecard

The Balanced Scorecard was developed by Robert S. Kaplan and David P. Norton. It was developed in a yearlong project that involved 12 companies in order to determine the best set of performance measures. The purpose of the scorecard was to provide managers with a quick but detailed summary of the performance of the company. In the Industrial era, the financial measures that were provided by a company worked well in the analysis of the performance of a company. However, these measures are not sufficient for companies of today. Top management understands that traditional measures such as return-on-investment and earnings-per-share may not be telling the full story. In order to counter this, some managers and researchers have focused mainly on the financial measures and others have focused mainly on measures that concern the operations of a facility. Kaplan and Norton emphasize that managers do not have to choose between financial and operational measures. The Balanced Scorecard shows the financial measures, which show the consequences of decisions, and it matches those measures with operational measures such as customer satisfaction, internal processes and the learning and growth of the company.

The Balanced Scorecard has many advantages but there are two that Kaplan and Norton highlight. Firstly, the balanced scorecard forces managers to focus on a set of the most critical performance measures of an organization. This means placing seemingly unrelated measures on a single management report. Secondly, the balanced scorecard assists in making sure that managers do not focus on one particular area but that they look at all of the important measures. This encourages managers to identify relationships between measures because focusing on one measure may mean sacrificing on another KPI even though managers are not aware (Norton and Kaplan, 1992).

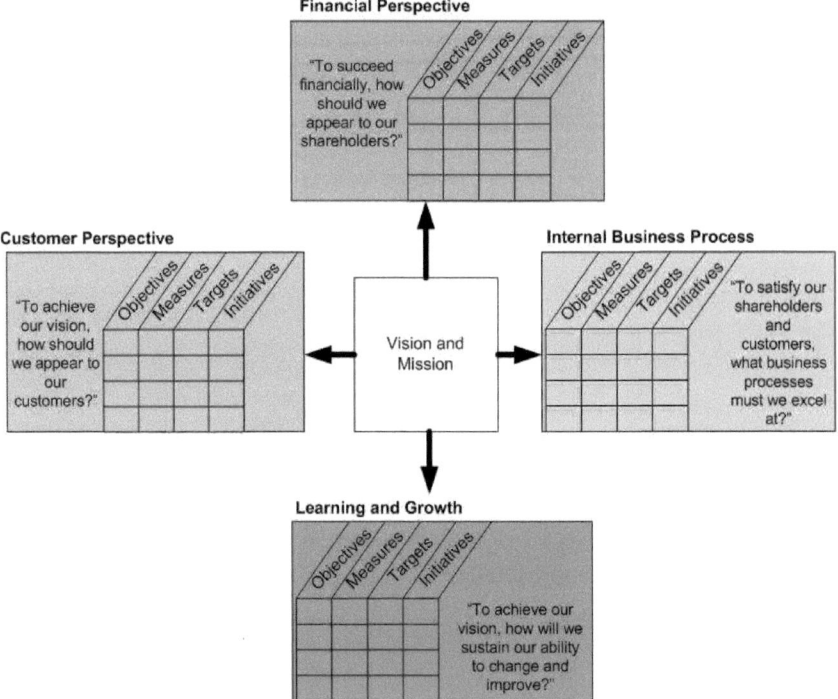

Figure 4: Balanced Scorecard Framework (Norton and Kaplan, 1996)

Figure 4 shows the four perspective of the Balanced Scorecard. Financial, customer, internal business process and learning and growth perspective all make up the Balanced Scorecard Framework.

7.2. Customer Perspective

In the customer perspective, managers decide on which customers to target and the market segment to contend in. Value proposition should also be defined, as in what is the value proposition being given to customers in a specific market segments. Typical customer measures include customer satisfaction, customer retention, new customer acquisition (Norton and Kaplan, 1996)

The balanced scorecard demands that an organization translate their mission statement into precise indicators that reflect the aspects that really matter to the consumer. The consumers concerns fall into the following 4 sections – time, quality, performance and service (Norton and Kaplan, 1992).

7.3. Internal Process Perspective

The Internal Process is a translation of how the consumer measures affect the internal processes. So the consumers measures help define the internal processes of an organization. It is important for the organization to focus on the internal process that satisfies the consumer. These aspects are usually – cycle time, quality, quality of staff and productivity (Norton and Kaplan, 1992).

7.4. Learning and Growth

This perspective is the area to focus on in order to provide competitive success. The goals for this perspective keep changing due to the severe global competition and thus call for a continuous improvement approach. An organization's ability to improve, learn and innovate is closely linked to its value (Norton and Kaplan, 1992).

The learning and growth perspective are based on three fundamental sources: people, systems and procedures. The financial, customer and internal process perspectives should reflect gaps of people, systems and procedures. The progress of closing that gap through educating employees, improving current technology/machinery etc is documented in the learning and growth perspective. Typical employee orientated measures include employee satisfaction, employee training and such (Norton and Kaplan, 1996).

7.5. Financial Perspective

Many senior managers have criticized financial measures due to the fact they focus on the past and they are unable to create value adding action. However, an organization's financial perspective is a measure of how the organization's plans, implementation and execution affect the finances of the organization. The measures of finance are the profitability, growth and shareholder value (Norton and Kaplan, 1992). The financial perspective has two parts – they outline the financial performance expectations based on the strategy and they are the final target for all the other perspectives (Norton and Kaplan, 1996).

7.6. Linking the Balanced Scorecard to the Company Strategy

The purpose of measurement systems such a Balanced Scorecard is to communicate a company's strategy. Robert Kaplan states that a Balanced Scorecard cannot be successful if it does not communicate the company's strategy. He goes on to say that those companies that show their strategy through their Balanced Scorecard are in a much better position to execute their strategy than companies who do not(Norton and Kaplan, 1996).

There are 3 main principles that are used to link a balanced scorecard to a strategy.

1. Cause-and-effect relationships
2. Performance Drivers
3. Linkage to financials

Cause-and-Effect Relationships

The strategy is a plan based on cause and effect. The cause and effect relationships are expressed through a series of if-then statements.

"If we increase employee training about products, then they will become more knowledgeable about the full range of products they can sell; if employees are more knowledgeable about products, then their sales will improve. If their sales effectiveness improves, then the average margins of the products they sell will increase."(Norton and Kaplan, 1996).

Outcomes and Performance Drivers

Outcome measures are the measures that display the previous performance of the organisation without an indication of how the organisation is performing with respect to achieving its current goals.

Performance Drivers are measures that are unique to the business strategy of the organisation. These are the measures that the organisations choose to compete in, in order to achieve their strategy. An example of this is measuring the profitability in the market section that the organisation has decided to complete in. Performance Drivers are lead indicators.

A good Balanced Scorecard should contain Outcome Measures (lagging indicators) and Performance Drivers (leading indicators)

Linkage to Financials

The Balanced Scorecard was created because financial measures were in themselves insufficient in showing an organisation's performance. Even if this be the case, the "bottom line" of financial performance should be emphasized in any Balanced Scorecard. The measures in the balanced scorecard should informally show the link to the financial intentions of the organisation(Norton and Kaplan, 1996).

8. Research

The objective for the research was to obtain all the information from the relevant sources in order to define Key Performance Indicators for the depot. Different views were considered when conducting the research. These views include the depot staff, customers in the form of fleet managers and private garages.

8.1. Pretoria Garage Personnel

Every organisation has performance standards. The standards may not be accurate or reliable but they are in place. This information lies with the technical advisors that are working in the garage. Technical advisors are the leaders on the garage floor. They are also known as the team leaders and each manages on average 5 artisans and apprentices.

The information gathered from the different mechanical repair and panel beating technicians is as follows.

8.1.1. Mechanical Repair:

In the mechanical workshop when a vehicle enters the bays, the technical advisors estimate the duration of the process, write that down on a white board then assign it to a mechanic. A "good" job is one that is completed within specific time limits and is considered "perfect" when it is not returned for a subsequent repair. The repair can take 1 – 7 days but this may be longer depending on how many technicians are working on it.

8.1.2. Panel Beating:

The same process that takes place in the mechanical workshop follows suit panel beating workshop in terms of there being a written time before the vehicle is assigned to a panel beater. A job is 'good' if it is done on time and properly.A panel beating job can take anything between 1 day and 3 weeks. They receive about 35 vehicles a day and complete 5 daily.

8.2. Fleet Managers

A fleet manager is the person who can be considered as the customer because they work at the police stations and are in charge of the vehicles. They facilitate the police schedules; document the monthly mileages for each car and general management of the fleet. They also communicate directly with the depot to make appointments, and check on the progress of their vehicles. It is important to know what they consider when deciding how good (or bad) a garage is.

8.2.1. Fleet Manager, Pretoria North Station

The Pretoria North fleet manager identified good service and quick turnover as key performance indicators of the depot. In addition he mentioned that the Pretoria Central Depot does not satisfy any of the above criteria. Cars usually take a month for mechanical repairs and panel beating at the Pretoria Central Depot.

8.2.2. Fleet Manager, Brooklyn Station

The Brooklyn fleet manager relayed that he appreciated that he could visit his vehicles at the Pretoria Central depot at any time to find out their progress and how when the vehicles would be repaired. However, he disliked the attitude of the receptionist who helped him as they were not very keen to help. The factors that this Fleet Manager considers when analysing a depot is the cycle time and how the customers are treated.

8.3 External Sources

External sources considered are private garages. These are important as they provide insight into industry and what is the standard for performance indicators is and how they are measured. These organisations work towards making a profit.

8.3.1. Private Garage 1

Private Garage 1 performs service and repairs. The services take 3 hours, "big repairs" take 1-2 weeks and if the repair takes too long, they then issue a courtesy vehicle (a vehicle that the customer may use whilst theirs is being repaired). In order to know what the customer thinks they observe the customer's body language, comments and facial expression about their product/service based on the customer's attitude when fetching the vehicle. They also have a customer care line where customers can call and air their views. Most of the complaints are about the car itself in terms of service, reliability and fuel consumption.

8.2.3. Private Garage 2

Private Garage 2 performs service and repairs only. Services and minor and major repairs are all done in one day. For customer feedback there is an independent company that is hired by Private Garage 2 which calls customers the day after their experience. The manager primarily checks how well the shop is working through a BSI (Business System Integrated automotive is an excel run programme that documents the progress of vehicles and other useful information) screen by looking at – "Cars in, cars out" and parts waiting. Communication with customers is as follows: an SMS is sent before the day of appointment, SMS is sent in workshop and SMS is sent in the wash bay and SMS is sent when the vehicle is ready to be fetched. This is done manually through the BSI system. This garage measures their mechanics according to the amount of hours sold. Every mechanic has a certain amount of hours available and they are paid according to the amount of hours they work.

8.2.4. Private Garage 3

Private Garage 3 performs mechanical repair only. They analyse their customer feedback by giving their customers a form to fill in. The manager would use the information from the customer satisfaction form and the BSI system to check the progress of the shop. The screen displays a lot of information with the, the manager's priorities being:

1. SMS notifications to customers
2. The number of courtesy vehicles available
3. Status, how long has a vehicle been in the shop.

8.2.5. Private Garage 4

Private Garage 4 processes panel beating only. This garage has no set performance indicators. Customers may go to the AA (Automobile Association) and check if their vehicles have been repaired to acceptable standards. If something has not been repaired properly then it can be brought back. Quality is their greatest performance measure. No set duration for panel beating. Minor repair take 1-2 days and major repairs takes 2 weeks – 1month.

8.2.6. Private Garage 5

Private Garage 5 does panel beating only. For this garage, time is their greatest factor. They have a customer survey for customer feedback. There is no set time for panel beating. Small paint jobs may take 2-3 days and major jobs take 8-10 working days.

9. Analysis and Recommendations based on the Research

Based on the research gathered the following performance indicators were compiled, accompanied by a recommended measurement method.

9.1. Time to Process

9.1.1. Repairs

The time it takes for a repair to be completed is very important. A garage creates its own expectation because its sets its own time standards. Private garages can usually complete a service within 2 hours, a small repair within a day and big repair in 1-2weeks. Time is important because it is a "commodity" that customers sacrifice when they bring their cars in to be repaired.

9.1.2. Panel Beating

After a vehicle has had an accident, there may be a need for it to go through panel beating in order for it to be fixed. This has the potential of being a lengthy process because it requires a large number of orders to replace the demolished parts and a number of processes. For this reason there is much less expectation for the cars to be repaired speedily and thus less emphasis on time in this particular area of service.

Recommended Measurement

<u>Macro</u>

From a high level, the number of cars that enter the depot is compared to the number of vehicles that leave the depot per period of time and are ready to leave the depot after their fleet manager has been notified.

<u>Micro</u>

This is tracking of the time it takes for service, repair or panel beating per vehicle. From the day they enter the facility until the day that the fleet manager is notified of the vehicles availability.

9.2. Quality of Process

The quality that is associated with a particular company is a direct reflection of how a company is able to do what it has set out to do. This refers to how well the depot can repair or panel beat a car.

Recommended Measurement

The percentage of vehicles that come to the depot for panel beating or repairs and leave the depot after having been repaired or panel beaten but have to return within a particular period of time (guarantee) because of either incomplete or incorrect work done by the depot.

9.3. Customer Feedback

This is the leading factor with regards to performance indicators as it is directly related to the income of a service/repair/panel beating of a private depot. This indicator will most likely alert the depot of other performance indicators and their performance.

Recommended Measurement

Customer Survey (see Appendix)
Particular areas that the survey addresses according to the questions:
1. Customer interaction: very bad, bad, good, excellent.
2. Time for service/repair/panel beating: very bad, bad, good, excellent.
3. Quality of the repair/service/panel beating: very bad, bad, good, excellent.

4. SMS Notifications: Yes or No
5. Return: Yes or No
6. Overall: very bad, bad, good, excellent.

	Excellent	Good	Bad	Very Bad	Yes	No	Weight
1	1	2	3	4			16%
2	1	2	3	4			16%
3	1	2	3	4			16%
4					1	2	16%
5					1	2	16%
6	1	2	3	4			20%
Total:							100%

Table 2: Table representing calculation of customer survey

Each question on the survey addresses an important aspect of the depot and finds the customer's perspective on it. From 1 to 6, the questions address the customer's opinion on the staff interaction, time, quality, whether they got notification or not, whether they would recommend the depot to others and their overall experience. Each question has a particular weight and once a rating (very bad, bad, good, excellent) or an option(yes or no) is picked, that percentage is chosen from that question. A combined total of between than 17%-34% is dismal as 17% is the lowest score one can have and 34% is the sum of the Bad scores. This shows that there is little, if any, that the customer appreciates about the depot. Between 34%-50% is still bad but there is one or two areas that are good but the bad percentages still outweigh the good ones. Between the rating of 50% and 66%, the performance of the depot is good. Between the rating of 64% and 100%, at least 4 of the 6 criteria are excellent and this is a good indication that the customer enjoyed their experience.

9.4. Employee Satisfaction

This is the measurement of the degree of satisfaction that is experience by employees throughout the depot. This indicator is important as it highlights the one of the most important contributor's views of the depot'. This indicator reflects an indirect contribution to the performance of the employees and their attitude.

Recommended Measurement
Employee Satisfaction Survey (see Appendix).

9.5. Employee Performance

This indicator measures the performance of an individual according to the repair or panel beating responsibilities that have been assigned

Current Measurement

The depot has a practice of assigning repair or panel beating jobs to certain individuals in that workshop throughout the month. As the month advances, feedback is given about the assignments and by the end of the month a percentage is calculated according to the assignments completed. Example: Kobus de Vries completed 85% of his assignments.

9.6. Employee Repair Capacity

Employee repair capacity refers to the number of employees that are available at the different service, repair and panel beating workshops. There is a certain amount of employees that are supposed to be at the workshops and if the number of employees does not reach the quota then it will reflect in the rate and the quality and/or quality of vehicle processing.

Recommended Measurement

The suggested measurement for this indicator is calculating the percentage of employees that are in the facility or workshop against the suggested quota.

9.7. Number of Outsourcing Jobs/Outsourcing Costs

Outsourcing or subcontracting is the practice of taking vehicles to a private garage to be fixed. This can happen for a number of reasons such as insufficient work bays, insufficient staff, not having the relevant equipment etc. A significant difference between performing a job at the depot and outsourcing it is the cost.

Recommended Measurement

Measure the monthly cost to outsource and the reason for outsourcing.

10. Strategy for the Depot

The main principles the Balanced Scorecard of Cause-and-Effect relationships, outcomes and performance drivers and ultimately linking KPIs to finances will be addressed in this section.

10.1. Cause-and-effect relationships

The strategy of the organisation is derived from the mission and vision of the depot which are:

Mission: To provide a cost-effective, professional and client-oriented mechanical maintenance service in order to meet operational needs in terms of transport in the South African Police Service.

Vision: To provide solutions in the transport environment with trained and dedicated personnel.

Figure 5 shows the recommended depot strategy based on the mission and vision of the organisation.

Decrease outsourcing costs and improve customer satisfaction strategy

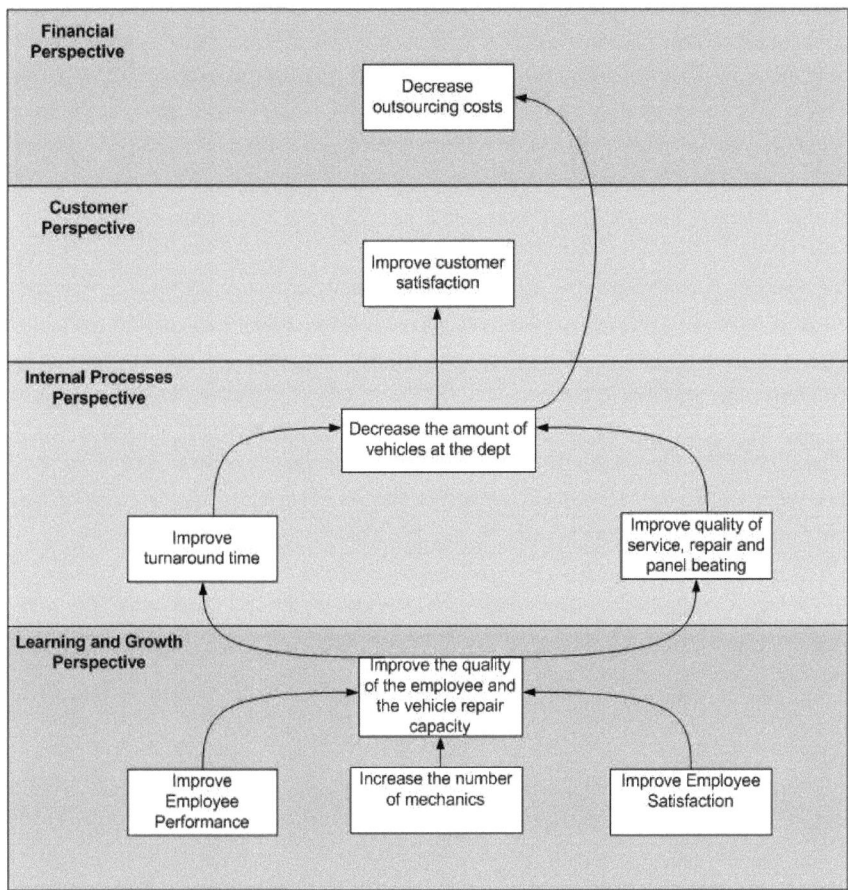

Figure 5 shows that within the learning and growth perspective one can see that the ultimately the goal is to develop the quality of the employee and the increase the vehicle repair capacity. Employee satisfaction can be improved from gaining feedback from employees about what would make their work experience better so that they may have a better experience at work which will positively influence their attitude towards their work and therefore improve the quality of work. The depot Colonel has stated that the station functions below the amount of staff that is supposed to be there, therefore increasing the amount of staff would assist the depot in their repair capacity.

From the improvement of the number of the employees, employee satisfaction and employee leadership, there will be a decrease in the amount of time that the vehicles spend at the depot (also known as turnaround time) and an increase in the quality of the repairs and panel beating. This will result in a decrease in the amount of vehicles in the depot. This is due to the low turnaround time and vehicles that do not return to the station due to low quality repairs.

The decrease in turnaround time and the increased quality of repairs will result in increased customer satisfaction. Unlike usual balanced scorecards, the customer perspective has a very weak relationship with the financial perspective. The fleet managers, who are considered to be customers, are essentially obligated to go to the Silverton depot for their vehicle's repairs due to the fact that the government pays for their repairs. If the turnaround time and the quality of repairs greatly deteriorated and the customers were very unhappy, it would be unlikely that they go elsewhere to service, repair or panel beat their vehicles. More importantly, any decision they make will not affect the depot's finances as they do not pay for any of the depot services.

The improved internal processes will also result in decreasing the cost to outsource. This is the case because some of the vehicles at the depot are outsourced due to overcrowding of the depot and insufficient staff to handle the repairs.

10.2 Outcome Measures and Performance Drivers

The following table shows the Outcome Measures (lag indicators) and the Performance Drivers (lead indicators) of the recommended Key Performance Indicators.

Key Performance Indicators	Lag	Lead
Employee Performance	Yes	
Employee Satisfaction		Yes
Employee Repair Capacity		Yes
Time	Yes	
Quality	Yes	
Number of Outsourcing Jobs/ Outsourcing Cost	Yes	

Table 3: Lag and Lead Indicators

10.3 Linking to Financials

Figure 5(on strategy) shows the link to finances. It shows that the decrease in the number of vehicles the depot will result in a decreased amount of outsourcing jobs and subsequently decreased outsourcing cost.

11.   The Balanced Scorecard

The balanced scorecard in this project will also be used as a measurement system for the depot's current performance. This is done by measuring its current performance and comparing it to where its aims to be.

In order to compute those calculations, the following information is needed.

Minimum: This is known as the lowest value for a KPI.

Maximum: This is the highest value for a KPI.

Baseline/Value: The baseline is the current value of the KPI.

Target value: The target value is the planned value for the KPI or where the depot aims to be in the future.

Weight: The weight of a KPI is the importance of a KPI relative to other KPIs.

Optimization: this refers to the goal of the KPI. Is the goal to maximize the value of the KPI such as in a customer satisfaction survey where one wants to maximize that value or is the goal to minimize the KPI such as the time it takes to perform a mechanical repair or panel beating?
The following formulas are used to calculate the relevant optimizations.

$$Maximization = \frac{Current\ Value(baseline) - Minimum\ value}{Maximum\ value - Minimum\ value} \times 100$$

$$Minimization = \frac{Maximum\ value - Current\ Value(baseline)}{Maximum\ value - Minimum\ value} \times 100$$

Performance: from the formulas of optimization the performance of the KPI is calculated.

Key Performance Indicator(units)	Min	Max	Baseline	Target	Weight	Optimization	Performance(%)
Outsourcing Costs							
Customer Survery(%)	17	100	78	66	1	Maxmization	73.49
Time: Mechanical Repairs(days)	1	21	6.64	7	3	Minimization	71.8
Time: Panel Beating(days)	1	21	16.53	10	2	Minimization	22.35
Quality: Mechanical Repairs(%)	0	10	6.89	0	3	Minimization	31.1
Quality Panel Beating(%)	0	10	0	0	3	Minimization	100
Employee Performance(%)	0	100	95.16	100	3	Maxmization	95.16
Employee Satisfaction(%)	0	100	59.91	66	3	Maxmization	59.91
Number of Mechanics(people)			77	73	3	Maxmization	100

Table 4: Summary of KPI Information

Note: the value for the time taken for the panel beating and mechanical repairs is not based on the initial definition. The information for this KPI should have been the time between a job card being opened (when the station takes responsibility for the vehicle) and when it is closed (after the vehicle has been serviced, repaired or panel beaten and the vehicle is ready to be fetched). However, the dates are based on when a job card was opened to the time that it was assigned to a particular workshop. This error is a result of the accounting system that is used to record vehicle information and because of the amount of time and paper required to reprint the information the error could not be rectified.

In terms of the weights of the KPIs, it is important to understand why the weights are distributed as displayed in Table 4 and in Figure 6. All of the KPIs are significant to the same degree with the exception of those that measure the time it takes for panel beating and customer satisfaction. Customer satisfaction counts for little in its contribution to the depot because it does not contribute in any way to the finances. Even if fleet managers are unsatisfied, it is unlikely that they will service, repair or panel beat the vehicles as they will have to pay for it. There is little emphasis on time for vehicles that are panel beaten because panel beating jobs generally take more time to complete. Fleet managers and customers to private garages expect that their vehicle will take a significant amount of time to fix.

From Table 4 one can see that the KPIs that are performing particularly well are the Customer Satisfaction, the quality of the panel beating and the number of mechanics at the depot. The performance indicators that are performing significantly negatively such as below (in the case of maximization) or above (in the case of minimization) their targets are the time for mechanical repairs, panel beating, the quality of mechanical repairs and employee satisfaction.

In table 4 also shows that there are no inputs for the outsourcing cost KPI. This is because there are too many factors that contribute to this particular KPI and that makes it particularly complex to create a target for it. The depot does not have an outsourcing budget. The cost of any repair or panel beating in-house or outsourcing is deducted from the book value of the vehicle. For example, a vehicle that is worth R66 000 may undergo panel beating that costs R16 000, after that repair the vehicle is worth R50 000. This example is meant to show that there are no specific budgets for the depot except in the form of the book value of the vehicles.

11.1. Balanced Scorecard Interface

Balanced Scorecard Designer is a software package that is specifically designed to communicate balanced scorecard measurements and strategies. This is very helpful in displaying the current status of the station.

Name	@	Performance		Value	Baseline	Target	Measure	Weight	Min	Max
⊟ Balanced Scorecard	○	**67.09 %**	→		**48.72**	**84.29**	**%**		**0**	**100**
⊟ ☆ Financial Perspective	○	50 %	→		0	100	%	3	0	100
○ Outsourcing Costs	○	50 %	→	50	0	100	Score	7	0	100
⊟ ☆ Customer Perspective	○	73.49 %	→		39.76	59.04	%	1	0	100
○ Customer Satisfaction Survey	○	73.49 %	→	78	50	66	%	1	17	100
⊟ ☆ Internal Process Perspective	○	63.11 %	→		63.11	71.5	%	3	0	100
○ Time: Mechanical Repairs	○	71.8 %	→	6.64	6.64	7	Days	3	1	21
○ Time: Panel Beating	○	22.35 %	→	16.53	16.53	10	Days	1	1	21
○ Quality: Mechanical Repairs	○	31.1 %	→	6.89	6.89	5	%	3	0	10
○ Quality: Panel Beating		100 %	→	0	0	0	$	3	0	10
⊟ ☆ Learning and Growth Perspective		86.04 %	→		86.04	89.8	%	3	0	100
○ Employee Performance		95.16 %	→	95.16	95.16	100	%	4	0	100
○ Employee Satisfaction Survey	○	59.91 %	→	59.91	59.91	66	%	3	0	100
○ Number of Mechanics		100 %	→	73	73	73	%	3	60	73

Figure 6: Strategy tree of the depot

Figure 7 is called a strategy tree. It gives us the overall performance of the depot which is based on the it's strategy. This diagram shows that even though the depot is working satisfactorily, there is still room for improvement.

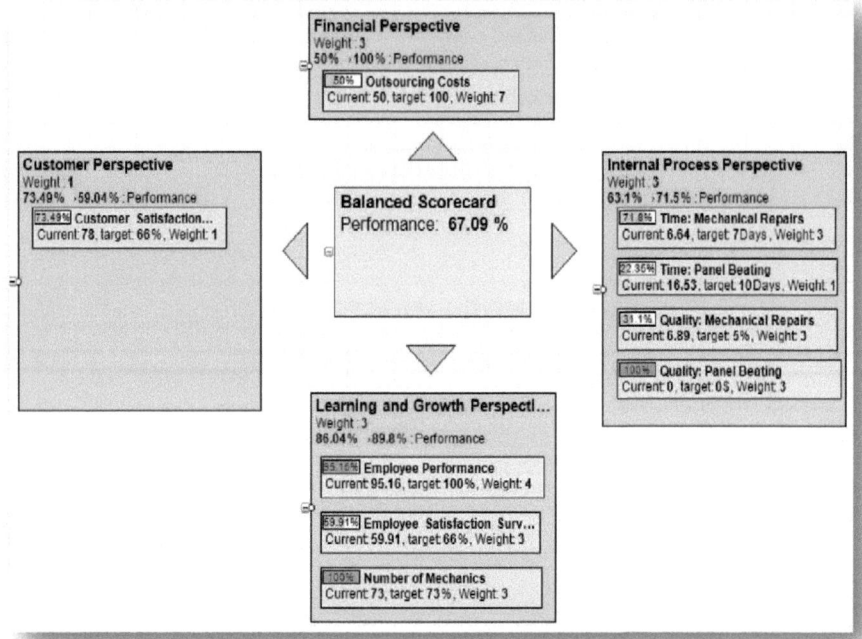

Financial Perspective
Weight : 3
50% →100% : Performance
| 50% | Outsourcing Costs
Current: 50, target: 100, Weight: 7

Customer Perspective
Weight : 1
73.49% →59.04% : Performance
| 73.49% | Customer Satisfaction...
Current: 78, target: 66%, Weight: 1

Balanced Scorecard
Performance: 67.09 %

Internal Process Perspective
Weight : 3
63.1% →71.5% : Performance
| 71.6% | Time: Mechanical Repairs
Current: 6.64, target: 7Days, Weight: 3

| 22.35% | Time: Panel Beating
Current: 16.53, target: 10Days, Weight: 1

| 31.1% | Quality: Mechanical Repairs
Current: 6.89, target: 5%, Weight: 3

| 100% | Quality: Panel Beating
Current: 0, target: 0S, Weight: 3

Learning and Growth Perspecti...
Weight : 3
86.04% →89.8% : Performance

| 95.16% | Employee Performance
Current: 95.16, target: 100%, Weight: 4

| 59.91% | Employee Satisfaction Surv...
Current: 59.91, target: 66%, Weight: 3

| 100% | Number of Mechanics
Current: 73, target: 73%, Weight: 3

Figure 7: Strategy Map

The diagram above describes the strategy map according to BSC designer software.

11.2 Balance Scorecard Dashboard

A dashboard is a summary of the performance of the depot. The dashboard and a strategy tree have a similar purpose of displaying the performance of the depot but they have different display features. The strategy tree is more comprehensive, detailed and has more writing than the dashboard whereas the dashboard usually employs more images such as graphs and pie charts and has less writing. Much like a speedometer, a Balanced Scorecard dashboard provides the viewer with enough information to know how well they are performing at first glance.

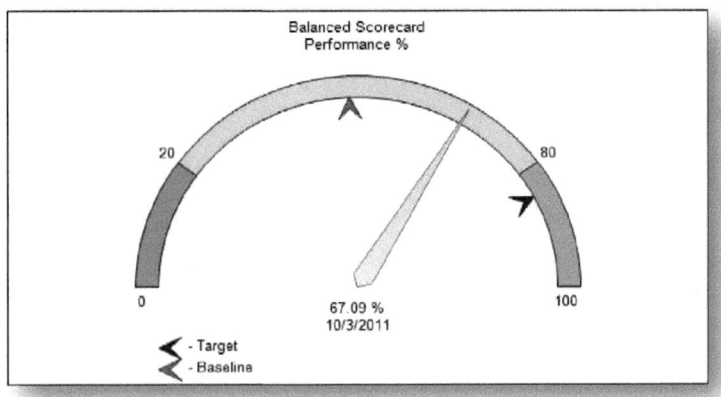

Figure 8:Performance Gauge

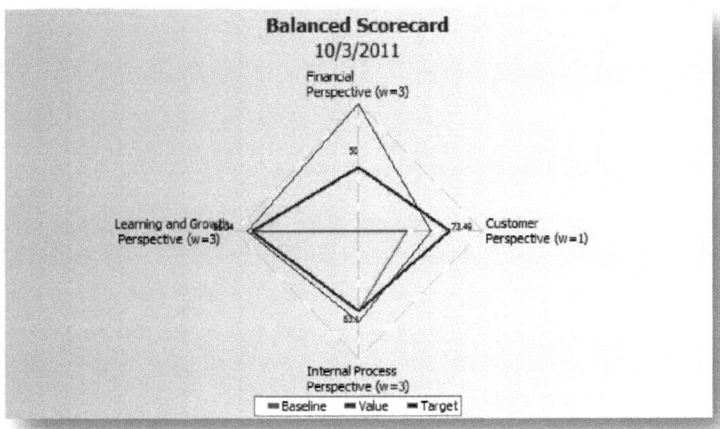

Figure 9: Diamond

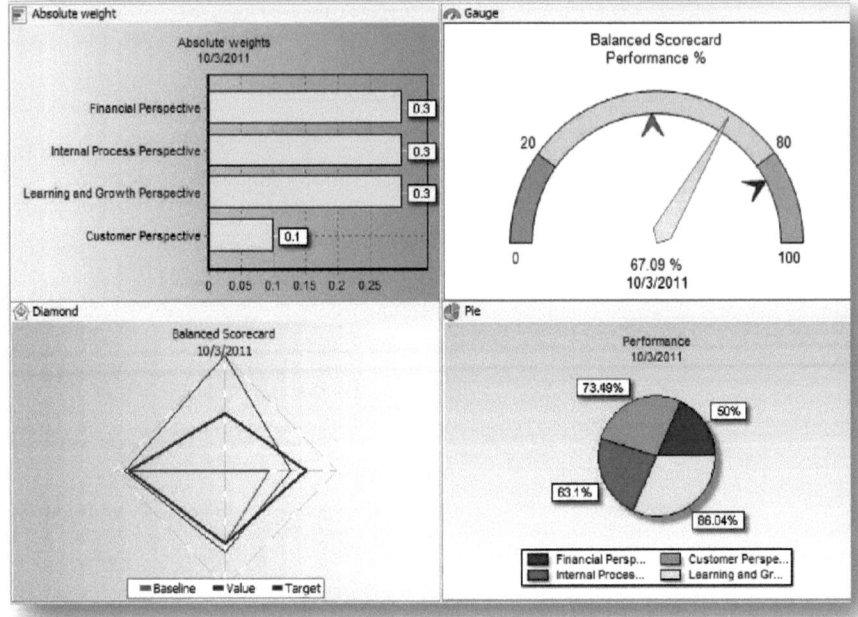

Figure 10: Dashboard

Note: for clarity, Figures 9 and 10 have been displayed independently.

The dashboard (Figure 10) contains a horizontal graph of the absolute weights of the perspectives, a performance gauge, a diamond and a pie chart. There are more features that can be used on in order to track the performance of the depot. The information in these figures conveniently displays the station's performance at one point in time but one can track their performance and progress over time through graphs or the many other features that the Balanced Scorecard Designer employs.

12. Key Performance Indicator Management

The Balanced Scorecard has shown that employees are one of the most important contributors to the performance of a facility. Figure 5 shows that all other contributors would be improved if the relevant ones related to the employee would be improved.

From the employee surveys, it has been gathered that one of the greatest causes for dissatisfaction is the fact that they work on a contractual basis. A contract worker refers to one that is not permanently employed. Their contracts are renewed on a regular basic and they are not afforded the benefits of permanent employees. Although they do work for the SAPS, they subscribe to an agency which is paid by the SAPS. Most of the employees from the depot are on contract. This job status causes employee concern because they have little job security. It is recommended that management should mould an open and honest relationship with the contract workers, giving them a platform to air their concerns. Contract workers often do not feel like "part of the team" so management's efforts to acknowledge them as such will go a long way in terms of employee satisfaction and thus improve other KPIs.

The current method of keeping track of vehicles is that a job card is opened on for a vehicle on the Workshop Accounting System (WAS) and when the job card is closed, it is closed on the same system. From the time that the job card is opened to the time that it is closed, it is recorded on hardcopy. This form of identification fulfils its duty but there are other more efficient ways to identify and keep track of vehicles.

Bar Coding is a system that is used to identify and recognise items with a particular code on them. A bar code is made of solid rectangular bars with white spaces in between. The construct of the bars and the spaces in between represents unique alphanumeric characters that are what identifies items. The application of bar coding as a means of inventory identification is mostly seen in grocery stores, at the cashier. Bar coding components include the Barcode itself, bar code printers and bar code readers.

Radio Frequency Identification (RFID) uses tags that have encoded chips in them. When the tag is in a specific range of a special antenna then the chip is deciphered by a tag reader. Radio frequency tags can be set or permanently encrypted and can be read from great distances away depending on the type of RFID used. There has been a big increase in the RFID technology used in recent years. RFID is mostly used in the materials handlings product identification, product tracking and in ocean shipping containers(Tompkins, J, White, J, Bozer, Y &Tanchoco).

13.   Conclusion

In closing, this document has shown the research that has gone into this project. This would be the process of getting information from the personnel at the depot, fleet managers and the private garages. This process was part of defining Key Performance Indicators by combining the views of the depot personnel, the customers and garages in the private sector.

After defining the KPIs, there is more focus on the Balanced Scorecard. The more intricate steps of the Balanced Scorecard begin to take shape when the mission and vision of the organization are used in order to shape the strategy of the depot. The strategy is what will guide the management of the depot in terms of what to focus on in terms of KPIs and why.

At this point the relevant statistics are incorporated into the document. The statistics are in Table 4. The calculation of min, max, targets, baselines etc are important to calculating the performance of the depot according to the Balanced Scorecard. The summary of the performance of the 4 perspectives in Figure 7 shows which perspectives are performing well and which ones need attention. The Balanced Scorecard Designer interface is an example of the interface that was mentioned in the deliverables.

Recommended tools such as the use of a dashboard system, bar coding and RFID technology are suggested as they will make for better inventory management and keeping track of KPIs.

This assignment has used a number of Industrial Engineering principles in order to achieve its goal of identifying KPIs, recommending manners of measurement, displaying the station's current performance through the Balanced Scorecard and making recommendations on those KPIs. The use of such tools helps in terms of making valuable information visible and thus easier to manage.

14. References

1. Burger J, 2009, "Emergency Response Services of the SAPS", in Gould C, *Criminal (In)justice in South Africa: A Civil Society Perspective*, Cape Town.

2. Kaplan R & Norton D 1996, *The Balanced Scorecard*, President and Fellows of Harvard College, United States of America.

3. Kaplan R & Norton D, (1992), "The Balanced Scorecard – Measures That Drive Performance", *Harvard Business Review*, January-February (No. 92105), p. 73, 75, 77.

4. *Minnaar A, Mistry D,* "Outsourcing and the South African Police Service", *Private Muscule*, p. 38.

5. Niven RP, (2002), *Balanced Scorecard Step-by-Step*, John Wiley & Sons, New York.

6. *SAPS Profile*, 27 July, 2011, http://www.saps.gov.za/saps_profile/history/history.htm.

7. Tompkins, J, White, J, Bozer, Y &Tanchoco, J 2010, Facilities Planning, 4thedn, John Wiley & Sons, United States of America.

8. U.S. Department of Transport 1999, *Asset Management Primer*, U.S. Department of Transport.

15. Summary of Definitions and Abbreviations

AA: Automible Association
KPI: Key Performance Indicator
SAPS: South African Police Service
BSI: Business Systems Integrated
RFID: Radio Frequency Identification
Autozone: A vendor that sells vehicle components within the depot.
Cluster: A section in a city/town that has been designated for certain police stations to patrol.
Depot/Garage: A garage that is used by the SAPS to service, repair and panel beat its vehicles.

16. Appendix

SAPS Fleet Management Project

In conjunction with: UP (University of Pretoria)

CSI (Customer Satisfaction Index)

Please complete the following 5 questions regarding you experience of the Depot/Garage

Please tick the appropriate block. Eg. ☑

1. How would you rate our staff in terms of how friendly and willing to help they are?
   - ☐ Excellent
   - ☐ Good
   - ☐ Bad
   - ☐ Very Bad

2. How would you rate the quality of the job done on your vehicle?
   - ☐ Excellent
   - ☐ Good
   - ☐ Bad
   - ☐ Very Bad

3. How would you rate the time it took to complete the required repair, service or panel beating?
   - ☐ Excellent
   - ☐ Good
   - ☐ Bad
   - ☐ Very Bad

4. Were you notified that your car was ready to be fetched or did you have to call to find out?
   - ☐ Yes
   - ☐ No

5. If we provided private services, would you bring your personal vehicle to our garage?
   - ☐ Yes
   - ☐ No

6. Overall, was our service satisfactory?
   - ☐ Excellent
   - ☐ Good
   - ☐ Bad
   - ☐ Very Bad

Additional comments:

SAPS Fleet Management Project

UP (University of Pretoria)

Please answer the following questions regarding your experience as an employee at the depot.

Employee Satisfaction Survey

1. Length of Service
☐ 0-3 years   ☐ 3-6 years   ☐ +6 years

2. My job worries/concerns you even when I am at home.
   ☐ Strongly agree
   ☐ Agree
   ☐ Disagree
   ☐ Strongly disagree

3. My job is challenging.
   ☐ Strongly agree
   ☐ Agree
   ☐ Disagree
   ☐ Strongly disagree

4. My Supervisor is a good leader.
   ☐ Strongly agree
   ☐ Agree
   ☐ Disagree
   ☐ Strongly disagree

5. I am proud of my job and its contribution to the SAPS.
   ☐ Strongly agree
   ☐ Agree
   ☐ Disagree
   ☐ Strongly disagree

6. The reward system of the SAPS is fair.
   ☐ Strongly agree
   ☐ Agree
   ☐ Disagree
   ☐ Strongly disagree

7. The garage provides me with the resources I need to do my job.
   ☐ Strongly agree
   ☐ Agree
   ☐ Disagree
   ☐ Strongly disagree

8. What I like most about working for the SAPS:

9. What I dislike about working for the SAPS: